LYME DISEASE IN
HORSES

A review of the history, facts, diagnostics,
therapeutics, and preventative measures of Lyme
disease in horses

Mark T. Reilly, DVM, Diplomate ABVP (Equine)
Plympton, MA

Lyme Disease in Horses
Copyright © 2019 by Mark Reilly

Library of Congress Control Number: 2019931771
ISBN-13:   Paperback: 978-1-64151-468-2
             PDF: 978-1-64151-469-9
             ePub: 978-1-64151-470-5
             Kindle: 978-1-64151-471-2

Printed in the United States of America

LitFire
PUBLISHING

LitFire LLC
1-800-511-9787
www.litfirepublishing.com
order@litfirepublishing.com

CONTENTS

LYME DISEASE IN HORSES 1

THE HISTORY OF LYME DISEASE IN HORSES 3

THE LIFE CYCLE OF THE TICK 9

DIAGNOSIS OF LYME DISEASE 17

TREATMENT OF LYME DISEASE IN HORSES 29

PREVENTION OF LYME DISEASE IN HORSES 35

LOOKING TO THE FUTURE 41

LYME DISEASE FACTS 43

OTHER TICK BORNE DISEASES 45

TREATMENT OF ANAPLASMOSIS AND EHRLICHIOSIS 49

PREVENTION OF OTHER TICK BORNE DISEASES 51

REFERENCES 53

ABOUT THE AUTHOR 55

LYME DISEASE IN HORSES

Lyme disease in my horse patients was (and is) a constant source of frustration for me and my clients. Working in an endemic area of this disease means the probability of seeing cases in some kind of routine basis. In my 25 years of practice, the diagnostic tests have changed, as well as how to best handle and treat a horse suffering from Lyme disease. Poor response to treatment, expensive treatments, and repeat infections were common.

I began to look for answers. Research done on Lyme disease in humans, horses, and dogs has been used to compile this book. I am happy to say we appear to have a much better understanding on the disease today and how to handle cases.

Please enjoy this collection of facts, research data, and experience. There will be better diagnostics, better preventatives, and a better understanding of case management in the future. For now, we are getting great results with the information contained in this book. Please do not hesitate to contact me with your questions or thoughts – Lyme disease is a hot topic for so many affected horses (and people!).

-Mark T. Reilly, DVM, Diplomate ABVP (Equine)

Lyme disease was first described in 1975 in 51 residents of Old Lyme, Lyme, and East Haddam CT. They were all diagnosed as having a unique form of multiple joint arthritis. In 1982, a PhD student, Willy Burgdorfer, discovered the organism responsible for the disease. The organism was found within a deer tick (Ixodes scapularis) on Long Island, NY. This organism is described as a slender, spirally undulating bacterium, now known as a spirochete. In 1983, the spirochete was named after their original discoverer, *Borrelia burgdorferi*.

Borrelia burgdorferi bacteria transmitted through the bite of a tick

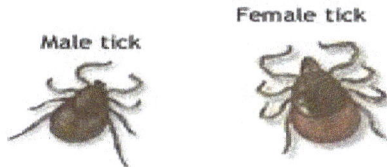

Male tick

Female tick

#ADAM

In 1984, the first case of dog Lyme arthritis was reported. The same signs – multiple intermittent joint arthritis, fever, and flu-like symptoms were all observed. As more medical professionals became aware of this disease, more patients were diagnosed from suffering from the infection.

In 2000, a quote in the veterinary medical literature stated Lyme disease "might not be the most prevalent equine disease you face with your horses, but it does exist, and can seriously impact a horse's health. Although documented cases of Lyme disease in horses are relatively rare, there are indications that the disease is on the increase."

Three years later, in 2003, research was being conducted at Cornell University on Lyme disease in horses. A research team led by Dr. Tom Divers reported 50% of horses in the Northeastern United States are positive for exposure to Lyme disease. This indicates at least 50% have been bitten by a tick carrying the Borelli burdoferi organism. Dr. Divers' team also reported success using intravenous oxytetracycline as a treatment for horses infected with Lyme disease. In addition, his research showed some protection against the disease in ponies vaccinated with a recombinant DNA canine vaccine; although safety and frequency of injections has not been determined.

One year later, a report stated "Up to 50% of adult horses may be infected" in the northeastern US. And by 2005, Lyme disease in horses had become commonplace. The veterinary literature is now quoted as saying, "Lyme disease is a problem more commonly thought to occur in our canine and human friends than horses. However, it does occur, and can have a wide range of signs and symptoms." And just one year later in 2006, spread of the disease into horses outside of the Northeast had become evident, marked by the statement, "As many as 20% of adult horses in certain areas of the US are infected with Borrelia burgdorferi. Many horses in endemic areas are, or have been, infected, which is evidenced by the fact that

75% of horses in the Northeast and Mid-Atlantic states already have antibodies against the organism."

The Center of Disease Control (CDC) reported that that Lyme disease accounted for 81% of all reported cases of arthropod-transmitted diseases in the US between 1986 and 1991. Lyme disease had now become the most common tick-borne disease in the US. Lyme disease currently accounts for more than 95% of all vector-borne diseases reported in the United States. There have been more than 128,000 cases reported since 1982. In 1998, the estimated incidence of Lyme disease was about 6 per 100,000 people in the U.S.; however, there may be considerable under reporting. In addition, incidence rates vary considerably from state to state and even within states and counties. In a few highly endemic counties, incidence rates exceed 100 per 100,000 people. State and local health departments can be consulted for more information regarding risk in particular areas.

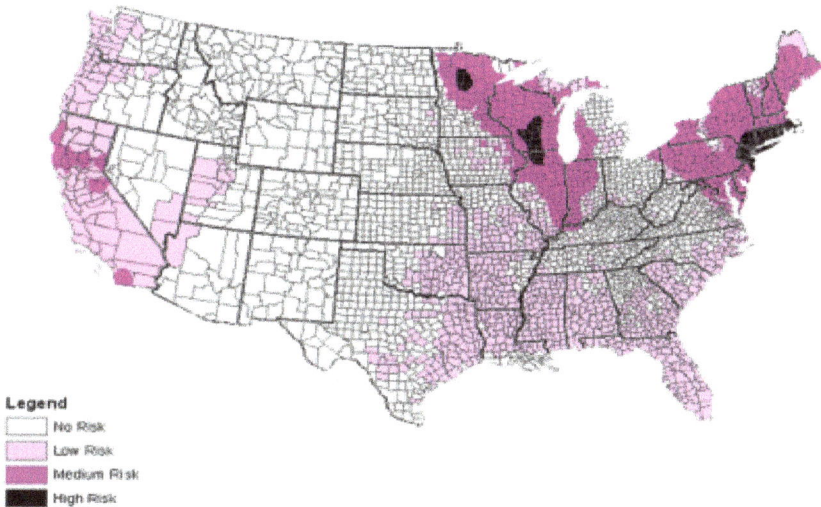

Legend
No Risk
Low Risk
Medium Risk
High Risk

Lyme disease distribution in the U.S.

According to data from the CDC, in 2014, 96% of confirmed human Lyme disease cases were reported from 14 states:

Connecticut	New Jersey
Delaware	New York
Maine	Pennsylvania
Maryland	Rhode Island
Massachusetts	Vermont
Minnesota	Virginia
New Hampshire	Wisconsin

Lyme disease is the most commonly reported vector borne illness in the United States. In 2014, it was the fifth most common Nationally Notifiable disease. However this disease does not occur nationwide and is concentrated heavily in the Northeast and upper Midwest.

The following two graphs have been taken from the CDC. The first one shows human Lyme disease patients are most likely to have illness onset in June, July, or August and less likely to have illness onset from December through March. The second graph shows the reported cases of human Lyme disease over a 20 year period (1995 - 2014).

Mark Reilly

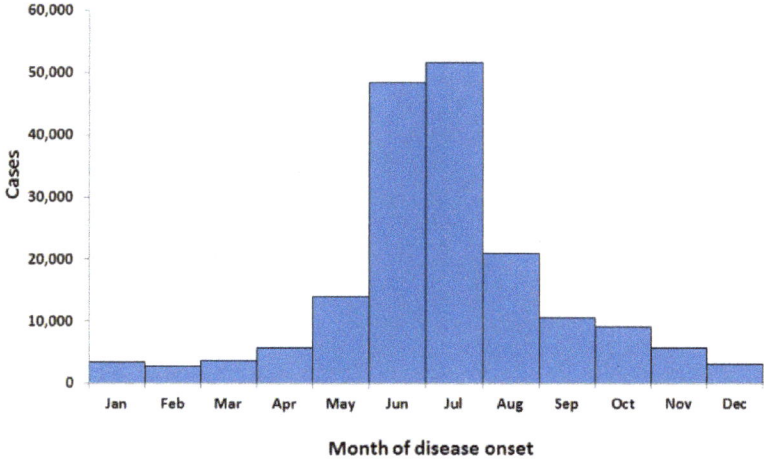

Confirmed Lyme disease cases by month of disease onset--United States, 2001-2010

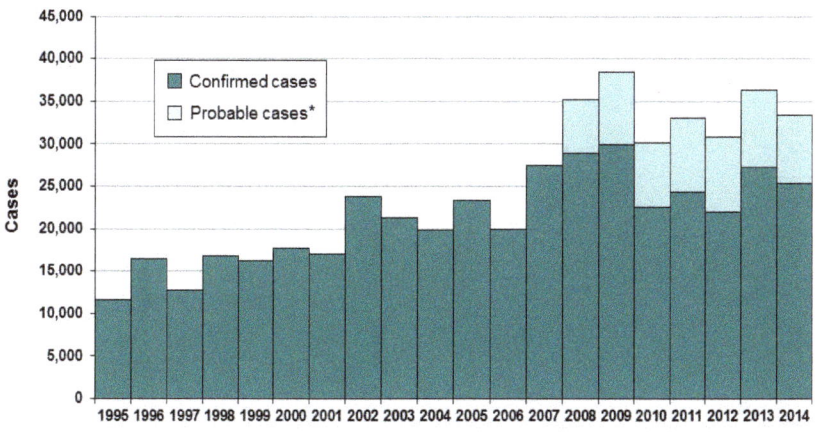

Reported Cases of Lyme disease by Year, United States, 1995-2014

THE LIFE CYCLE OF THE TICK

Ticks in Ixodes classification have 35 different species in North America and 250 species worldwide. The ticks are capable of transmitting viruses, bacteria, rickettsia, protozoans, and nematodes (Borrelia). These are each different types of organisms, some able to live on their own and others requiring other life forms to stay alive. They each cause different types of infection from different types of organisms.

Female & male adult deer ticks

The deer tick (Ixodes scapularis) is also commonly called a black legged tick. It has a requirement for three different hosts to complete its life cycle from egg to adult. Each feeding stage requires one vertebrate blood meal for its development. The adult female lays the EGG which then becomes a LARVA. The Larval stage now must complete a blood meal to become the next stage, a NYMPH. The nymphal stage must also complete a blood meal before changing into an ADULT. Each of the three feeding stages (Larval, Nymph, Adult ticks) is capable of causing infection. As the tick grows in size with each life stage it will feed on larger hosts. Humans and horses are likely hosts for adult ticks or nymphal stage ticks.

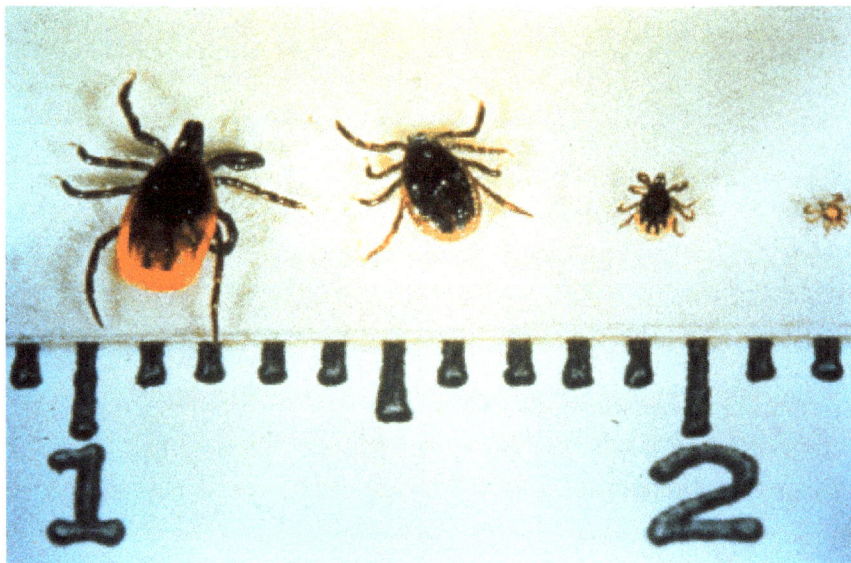

Horses are more likely to have a higher prevalence of Lyme disease than humans because ticks stay on horses longer. Research has shown that ticks must be attached for at least 24 hours to transmit their disease. The tick holds the organism in its simplified digestive system, known as a hindgut. A tick attaches to a host with its very sharp mouthparts. It then salivates and regurgitates into the host an anti-clotting agent so it can get a ready supply of host blood. While the tick is attached to the host it uses its mouth parts to suck and ingest blood of the host. It then sends this blood to its hindgut where it mixes with the organism. The tick then regurgitates back into the host as it continues to feed. If the tick had the Borrelia organism in its hindgut, it has now been deposited in the host. Not only does the host become infected, but is now serves as reservoir of Borrelia for the next tick to ingest when it attaches and continues to spread the disease.

The deer tick, or blacklegged tick, is commonly found in Northeastern and Midwestern US. Specifically, it is now commonly found outside of the northeastern U.S. in the states of Minnesota and Wisconsin. The ticks usually mate in the fall or early spring and will take two (2) years to complete their life cycle. While on a host (such as the white-tailed deer), the female engorges on blood. While she is engorging on her blood meal, the male mates repeatedly her and other females. The first breeding season occurs in the fall after the spring nymphs emerge from the larval stage in early fall. The now engorged female drops off the host and lives in leaves/brush or other protected areas until she lays her eggs. The female will lay up to 3,000 eggs at one time. The adult female then dies....

Engorged female

If, however, she did not feed in the fall and she overwinters and does not engorge until spring OR the eggs laid in the fall do not emerge as nymphs until spring, the second breeding season is now available in the spring. Therefore, there is a heavy distribution of larvae in spring and late summer when they emerge from the egg stage.

Eggs emerge to become larvae soon after being laid. The larvae are very small, about the size of a grain of sand. It is the 6- legged stage of the life cycle. Larvae need high humidity to survive; making spring and fall seasons ideal. Larvae are rarely infected, but must find a host to feed. This can be the beginning of infection. Larvae typically feed on small mammals for 3 – 5 days before dropping off and will metamorphosize to nymphal stages in leaf litter. The white-footed mouse population is full of mice infected with the Borrelia organism. It is estimated that 25% of white tail mice are infected. This makes the white-footed mouse the usual host for the cycle and continuation of Lyme disease. The tick has multiple opportunities to acquire the organism; as a larval or a nymphal stage. This increases the likelihood of a tick already infected with Lyme disease when it bites our horses as a nymph or an adult form.

The larvae and nymphs are known to infest 31 mammalian and 49 bird species. Adults have been found on at least 13 species of medium to large sized mammals. Back in 1994, 55-60% of ticks in New England were reported to be carrying the organism; we now have much better statistics which will be detailed below.

Larvae will metamorphasize into nymphs, now the size of a poppy seed. The 8-legged nymphs are most active in May, June, July, August. Larval activity is seen in May from fall eggs. The second and much larger larval activity peaks in August from the more successful spring breeding season. If infected as a larval stage; the infection is now transmitted to the nymphal stage. The nymph now seeks a host for a blood meal. This host is normally a mouse or smaller mammal, or bird. As it is the size of a poppy seed, it is easily undetected. But the tick now has the second opportunity to get infected from an infected host; get blood meal, drop off and turn into sexually active mature blacklegged ticks.

1: larvae 2: nymph 3: adult male 4: adult female

The adult stage of the deer tick is active fall to spring when the temperature is above 40 degrees. Adults feed on large mammals (horses, dogs, deer and human). Deer are reservoirs for adult ticks, but not for the organism itself. This means the deer do not have the disease in a high population, but they do act as a great host for ticks and can now spread ticks throughout a wide area with their movements. After engorging, the female tick drops off in a place to lay her eggs and the life cycle continues. As stated previously, an infected nymph passes on to an adult; now the 3rd carrier. Typically this is the only stage we find; if found and removed before 24 hours transmission can be prevented. The nymphal stage has the greatest Lyme disease transmission as the season is the longest with the highest number of hosts available (May – June).

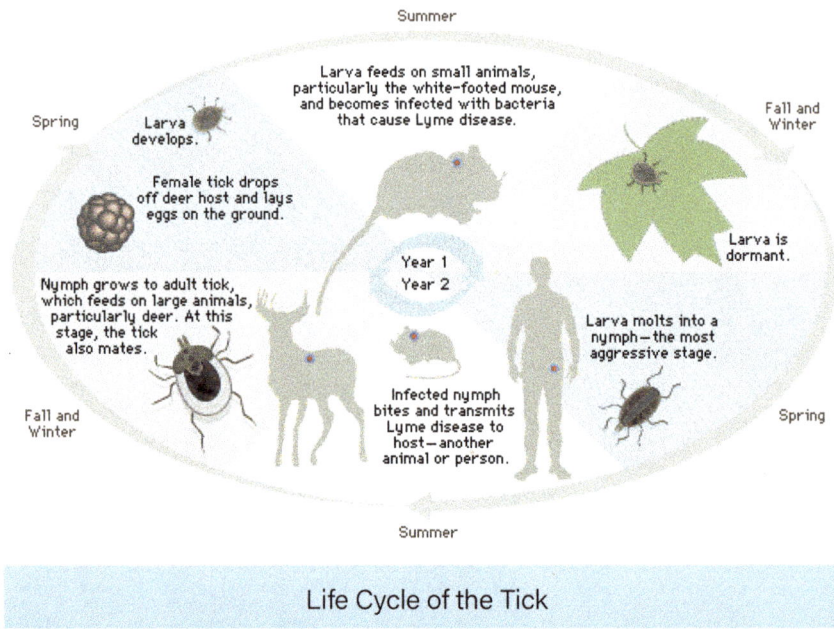

Summer

Larva feeds on small animals, particularly the white-footed mouse, and becomes infected with bacteria that cause Lyme disease.

Spring

Larva develops.

Female tick drops off deer host and lays eggs on the ground.

Fall and Winter

Larva is dormant.

Nymph grows to adult tick, which feeds on large animals, particularly deer. At this stage, the tick also mates.

Year 1
Year 2

Larva molts into a nymph—the most aggressive stage.

Fall and Winter

Infected nymph bites and transmits Lyme disease to host—another animal or person.

Spring

Summer

Life Cycle of the Tick

Earlier we explained the spirochete is transmitted to host via salivation and/or regurgitation. The spirochete is primarily found in the hind gut of tick, but it can be in salivary glands of the tick as well. Studies have shown it takes a minimum of 24 hours of tick attachment for spirochete transmission to the host. Due to the fact we can now see this adult life stage, and it takes some time to pass on its disease, we can interrupt transmission of the disease before it occurs. In addition as adults are most active fall to spring when temperatures are over 40 degrees, activity is often decreased in winter climates. Additionally, horses and humans are less active and have less exposure time as the weather cools down. Therefore, disease transmission decreases decidedly during adult season, even though up to 50% of adult ticks may be infected as we often remove them and the weather patterns are in our favor.

Making the diagnosis of Lyme disease can be challenging. The disease mimics many other diseases.

Fever, muscle aches, fatigue are similar to signs seen when a horse is infected with any number of viruses. Joint pain can be attributed to arthritis or soft tissue injury. Neurologic signs (weakness and incoordination) may mimic many other diseases such as Equine Protozoal Myelitis (EPM), wobbler's syndrome, cauda equine, or even encephalitis.

In humans, physicians often use a local reaction know as erythema migrans (EM) to diagnose a tick borne disease. This is a red, circular patch which appears 3 -30 days after the bite of infected tick. Initially, the reaction occurs at the tick bit site, but may then expand to cover a large area. It is typically non-painful. Unfortunately, finding EM is near impossible in our dogs and horses.

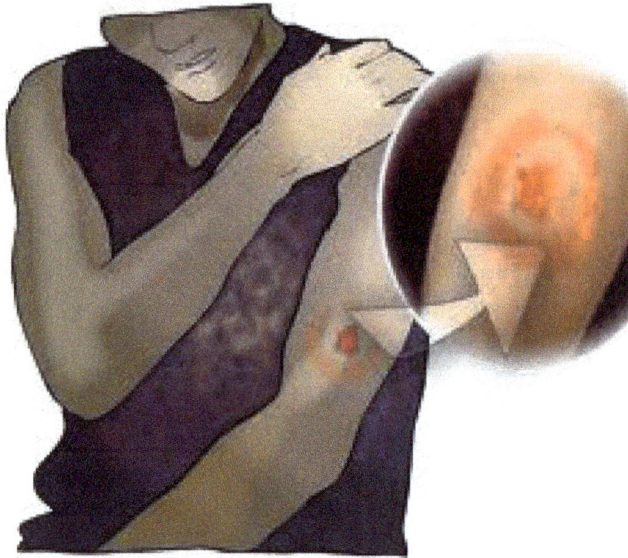

Erythema migrans "bull's eye"

***Human cases parallel animal cases:

    1982 – 523 reported cases
    1988 – 4,507 reported cases
    1989 – 8,552 cases
    1994 – 13, 043 cases
    2005 – 23,300 cases = 7.9 cases per 100,000 nationally (31.6 per 100,000 in the ten states where the infection is most common)

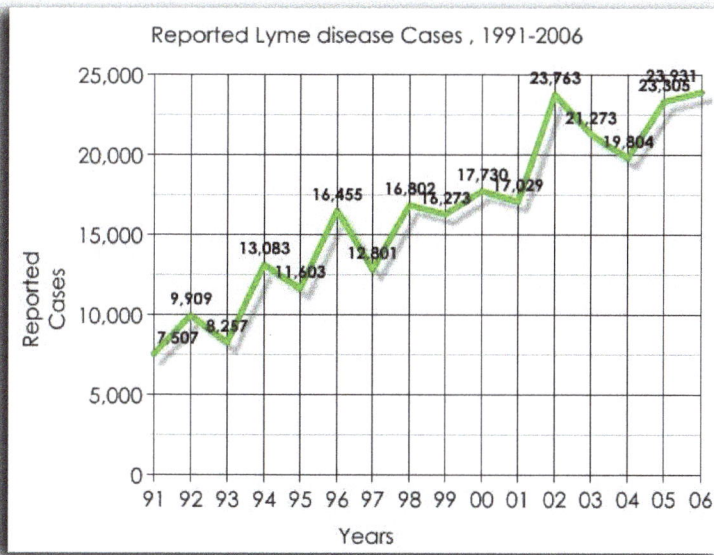

Reported Lyme disease Cases, 1991-2006

The US Dept of Health and Human Services has included Lyme disease among its prevention priorities. The Center for Disease Control (CDC) stated in 2005 that it had a goal of reducing the overall incidence by more than 40% in endemic areas by 2010. This goal has not been accomplished, although strides have been undertaken which have shown some success.

Below are more recent CDC state specifics on human cases which cases continue to increase:

Vermont
    2005 – 54   2012 – 386
    2006 – 105   2013 – 674
New Hampshire
    2005 – 265   2012 – 1002
    2006 – 617   2013 - 1324
Massachusetts
    2005 – 2336   2012 – 3396
    2006 – 1432   2013 – 3816

Connecticut
    2005 – 1810   2012 – 1653
    2006 – 1788   2013 – 2111
New Jersey
    2005 – 3363   2012 – 2732
    2006 – 2432   2013 - 2785

The best way to have Lyme disease diagnosed is to save the tick and have it tested for the presence of the bacteria. The UCONN Diagnostic Lab or Tick Research Lab at URI both offer this testing. Infection rate differs with geography: 50% of ticks in the northeast and the Midwest are infected, whereas only 1-2% of the ticks in the South & West test positive. Therefore, the chances of your horse being infected with Lyme disease may be 25 times greater here than in other parts of the country.

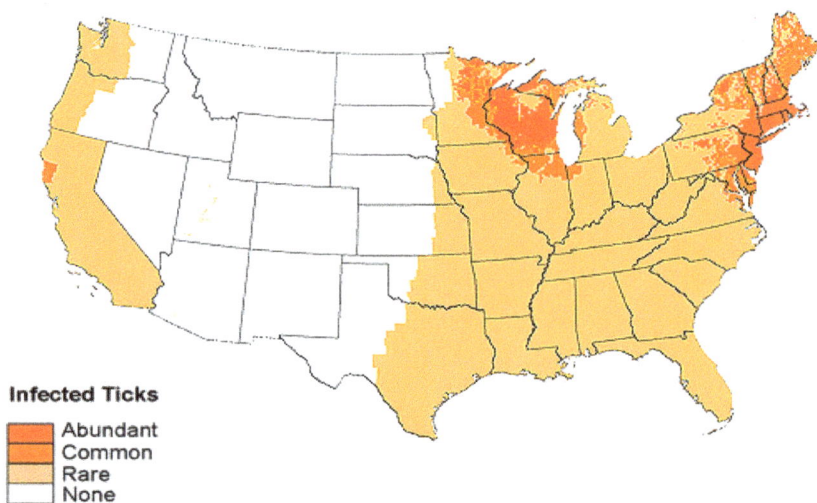

Infected Ticks

Abundant
Common
Rare
None

The disease is now found in 48 states, Canada, Europe, Asia, Africa, Japan, and Australia.

Blacklegged ticks are most abundant in rural locales with heavily wooded areas being infested in many countries. Lawns, gardens, bushes near houses are also likely to be infested in many areas.

Infected horses often have vague and variable signs. We see multiple different signs, sometimes more than one at a time. For example, signs most often reported include:

— stiffness/lameness

— muscle tenderness

— hyperesthesia (increased or altered sensitivity to sensory stimuli)

— resentment of touch or pressure

— swollen joints (rare)

— behavioral changes

  • Fever and edema are unlikely (more likely a result of Anaplasma phagocytophila infection, or a dual infection)

  • Unwillingness to work, lethargic, "grumpy"

We often have to rule out joint problems (osteoarthritis), muscle soreness or "tying up" injuries (rhabdomyolysis), neurologic issues (EPM), and swollen joint issues (OCD fragments). The diagnosis is difficult due to fact the most common presenting complaint is lameness, which may be intermittent and non specific. It may take a few days to weeks to develop clinical signs. There is not a specific type of lameness and there are numerous causes for lameness complaints. The lameness may be episodic or recurring and chronic. It can be the forelimbs, the pelvis, the spine, or the rear limbs. Central nervous system problems (encephalitis) have also been described. Eye problems described as similar to moon blindness (recurrent uveitis) have also been described. On the bright side: heart, liver, kidney problems have not been reported in horses, as they have in dogs and humans, and Lyme disease is rarely fatal in any species.

Just as in human, the ideal diagnosis is by testing the tick itself. This is rarely done, as quite often owners report the tick bite occurred weeks or months previously, or that a tick has never been seen. Therefore, we use a variety of blood tests.

A few words of vocabulary may help to understand the discussion on what the blood tests measure:

Antibody – a protein normally present in the body or produced in response to an antigen. The neutralizing of antigens by antibodies is called the *immune response.*

Antigens

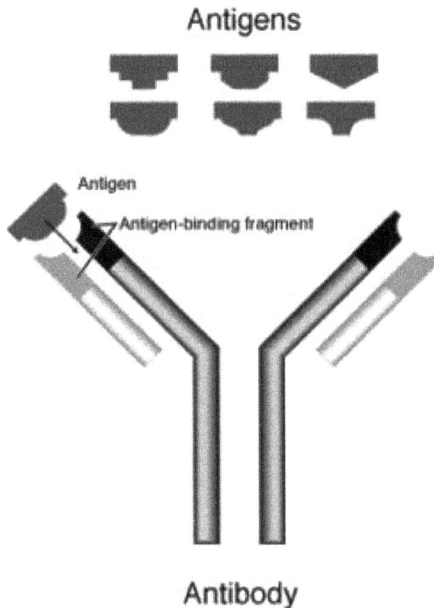

Antibody

Antigen – (bacteria) introduced into the body by the tick as it feeds. When they are detected, the immune system begins to create antibodies to fight the intruders.

Antigen binding fragment – at tip of antibody which identifies and binds (neutralizes) antigens

IFA or ELISA Test ("titer"): This test is performed by applying the patient's serum to an antigen-coated slide and adding an anti-body conjugate. The slide and conjugate are made by different companies (several different manufacturers), which affects the test dramatically. Different labs use different dilutional schemes for their titers. Interpretation of the test is technician dependent and read as a number of dilutions. A higher dilution indicates more antibodies. Unfortunately, titers do not equate to clinical signs or degree. Titers don't equate into degree of infection. This test measures the amount of IgG an animal has against Lyme disease. This means if an antibody test is done, a second test is necessary two weeks later to see if the titer is rising, falling, or not changing at all.

The animal's immune surveillance is possibly the reason why the animal will be positive on the serum antibody test but have no clinical signs. Ideally, one would always retest in 2 weeks and compare both titers to check for rising levels to positively diagnose a recently infected horse. If the test is reported as a high titer it can be interpreted as an exposed animal, an animal who has mounted a significant immune response, an infected and sick animal, or a recently cleared infection. If the test is reported as low, this can be interpreted as not infected, a cleared infection, or an animal that has not yet mounted an immune response. Therefore, this test lacks specificity and has many false positives (animals testing positive who do NOT have the disease). As stated earlier, 75% of horses in the Northeast have titers; therefore, we rarely perform an antibody test now that we have better choices.

Other tests commonly utilized include:

- Western blot (4 of 4 bands) – the "gold standard" takes several days (4) for the test to be completed. One must also transport samples to reference lab, and the results again can be ambiguous and requires repeating, especially within the first few weeks after infection. This test is also quite expensive over $100).

- Whole cell diagnostic methods (Kela units) measure the IgG antibodies produced to numerous antigens on the "whole" spirochete. The persistence of the whole cell antigens produce antibodies that are present and remain elevated even when spirochetes are reduced or eliminated.

- Lyme Multiplex Test – recent test formulated and utilized by Cornell (NY State Veterinary Diagnostic Lab). This luminex test looks at outer surface proteins A, C, and F (OspA, OspC, and OspF). OspA is important for localization of the organism within the midgut of the tick and is typically not expressed by the organism when it is in the mammalian host, but it is often found in horses, so it is unsure what its relevance is in the horse. OspC is a lipoprotein of the Borrelia organism which is abundantly expressed in mammalian hosts when the tick takes its blood meal. Expression of OspC facilitates movement of the organism from the midgut of the tick to tissues of the host. The early host immune response is primarily directed against OspC. As the host antibody titers to this protein increase, the organism must turn off expression of OspC or risk elimination from the host. OspC expression is greatly reduced by 10 days post-infection, and the antibody response typically tapers off by 7 or 8 weeks post-infection. OspF is generally expressed 4-6 weeks after the organism enters the mammalian host with resulting antibody response that occurs 6-9 weeks post –infection.

- SNAP ELISA test = C6 antibody. Introduced in 2004 as an in house dog test. The C6 antibody is associated with one of the invariable regions (IR6) of a surface protein of the organism known as VlsE protein. This VlsE gene is only expressed in the mammalian host; it is not expressed when the organism is within the tick or when grown in culture to produce the Lyme vaccine. Therefore, antibodies to the C6 peptide are an indication of natural infection with B. burgdorferi.

- Experimental infection studies have demonstrated that the higher the C6 antibody levels, the greater the number of organisms that could be recovered from the skin or tissues of infected animals. Furthermore, these studies have shown that these organisms are more likely to survive in culture when removed. Concentrations of C6 antibody decline rapidly in response to antibiotic therapy and so do the numbers of organisms that can be recovered from the infected animal. This means the test declines rapidly and significantly after effective treatment has removed the organism, as the previously mentioned invariable region would no longer be expressed. This test will detect antibodies against Lyme disease as early as three (3) weeks post infection and has shown to be 100% consistent with Western Blot results.

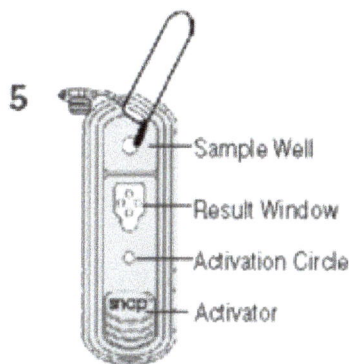

4DX SNAP Test

The Lyme disease SNAP test is comparable or superior to the traditional 2-tiered testing of IFA and Western Blot. The fact that a decrease in the amount of antibodies against C6 can be seen quickly will indicate a successful therapeutic outcome for Lyme patients. A drop of 50% or greater indicates a successful treatment. In 2005, a study validated this test in horses. A study reported in 2006 showed this test to have a high specificity and sensitivity (rare false positives or false negatives). This is a quick, reliable, in-house test, has been validated for horses, and *costs at least half of a Western Blot assay.*

The 4DX SNAP test takes 8 minutes and uses 3 drops of blood. It can be done **stall side**.

A study measured 164 horse samples – 106 positive with Western blot, 109 positive with SNAP (all 3 tested positive with IFA) = 99.4% sensitivity (it did not miss any positives, but identified a few more which may or may not be true positives)

A repeated test 60 – 90 days after completion of antibiotics of therapy can be used to assess success of treatment and status of the horse.

Notes on the SNAP test: Any color development indicates a positive result. You should not draw a correlation b/w the color intensity of the sample spot and the level of infection. It takes 8 minutes to read final result; waiting longer does not increase the accuracy of the test.

The test actually has four different tests, known as the 4DX. It is most commonly used in small animal practice to test dogs for Lyme disease, Heartworm disease, Anaplasmosis and Erhlichiosis. The Lyme portion identifies infection by measuring the C6 peptide which is only made by the host when the live organism is present within the host. The C6 peptide is highly specific to Borrelia burgdorferi and is <u>only present in the face of active infection</u>. There is also no cross reaction from vaccines. It is extremely accurate, having a sensitivity of 99.4% and a specificity of 100% for the Lyme disease portion of the test.

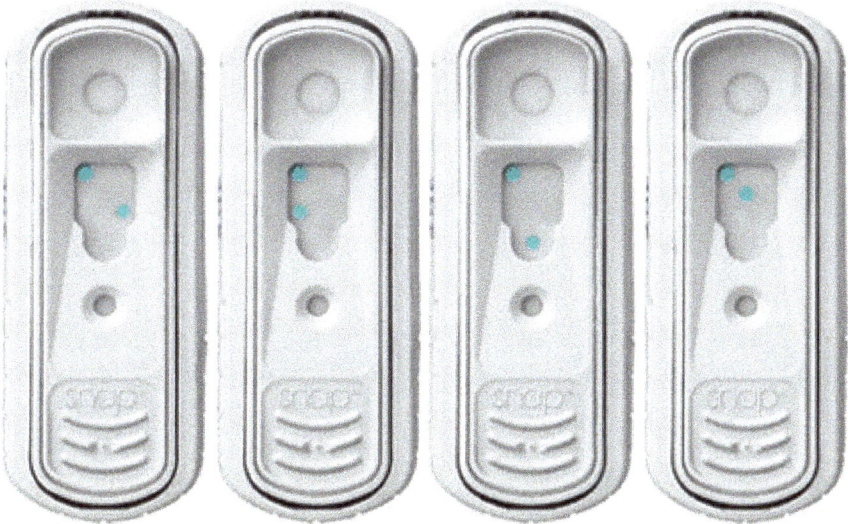

Heartworm	E. canis	Lyme	A. phagocytophilum
Sensitivity: 99%	Sensitivity: 99%	Sensitivity: 96%	Sensitivity: 99%
Specificity: 100%	Specificity: 100%	Specificity: 100%	Specificity: 100%
(95% CL 95-100%)	(95% CL 98-100%)	(95% CL 98-100%)	(95% CL 98-100%)

4 DX SNAP Test

As this is a bacterial infection, antibiotics are the basis of treatment. Numerous different antibiotics have been used to treat Lyme disease with varying reports of success. As reported by veterinary researchers at Cornell in 2005, **100%** of infected horses in one study were completely cleared of infection by using **intravenous oxytetracycline**. The study also compared two other antibiotics, doxycycline and ceftiofur (Naxcel); only to find that both were *only 50%* effective at removing the infection. We also know that doxycycline is poorly absorbed by the horse, and extended treatments may be necessary in order to utilize the horse's immune system to rid the disease while the doxycycline keeps the infection in check. In the study at Cornell, re-infection was common in both the doxycycline group and Naxcel group, but not in the oxytetracycline group.

Oxytetracycline

Minocycline is the latest member of the tetracycline family being offered as an oral treatment against tick borne infections, including Lyme disease. There is some research that suggests that oral minocycline may have superior bioavailability and reach higher tissue-concentrations in horses when compared to oral doxycycline. The dose offered is 4mg/kg twice daily for 3-4 weeks. Although there are anecdotal reports of successful treatment, there have been no pharmokinetic studies done in the horse to assess its bioavailability and success in eradicating Lyme disease. Personal communication with clinicians at UPENN New Bolton Center indicates that minocycline does not appear to offer the treatment success it was hoped to bring.

Body soreness, muscle soreness, and/or joint pain can be controlled with non steroidal anti-inflammatory agents (NSAIDs), such as Bute, Banamine, MSM and the like. Often treatment with NSAIDs is only necessary for 3 to 5 days while the antibiotic is mounting its attack against the organism. In addition, the joints that are inflamed are at risk for developing chronic problems or cartilage issues from the inflammatory mediators within the joints. Therefore, cartilage protective agents (chondroprotection) are often used simultaneously for the best outcome in many cases. These include *Cosequin*, or other oral products containing chondroitin sulfate and glucosamine hydrochloride. Injectable *Adequan* had been proven to repair damaged cartilage and may be prudent to use in competitive and/or athletic patients.

Acupuncture has also shown to be helpful in cases where attitude and body soreness are the major complaints. In addition, TCVM Herbals offer Lyme Formula to aid in the treatment of Lyme disease. Eastern medicine views Lyme disease being caused by invasion of pathogenic Wind-Damp Toxin when the body's Wei Qi (Defensive Qi) is too weak to defend against pathogens. This invasion leads to Qi-Blood Stagnation and therefore body soreness. This herbal acts to boost Wei Qi, clear Wind, drain Damp and resolve Stagnation. Lyme Formula is typically given for 3-6 months.

Recently, Resveratrol has become popular as an adjunct to helping relieve the discomfort of joint disease. Resveratrol is an antioxidant-rich plant polyphenol which protects the horse from the damaging effects of free radicals and inflammatory health problems, and offers immune support. This compound is found in red wine, as well as dozens of plants and in red wine. The author has used the product *Resvantage*, which sources its resveratrol from knotweed.

Many horses benefit from the addition of a probiotic while being treated with the antibiotics. The weeks of antibiotic treatment can alter the normal gut flora. Many probiotics will re-populate the intestinal tract and prevent potentially severe cases of colitis and diarrhea.

The Cornell studies involved using intravenous oxytetracycline at a dose of 6.6 mg/kg twice daily for 3 weeks in research ponies, kept in a disease free environment for the entire study. This is often not an option for numerous reasons – economics to the horse owner, compliance of the horse to twice daily injections, or the necessity of removing the horse from work for an extended period of time.

Therefore, we have used the information and research available to us to make a successful protocol which does not inconvenience the horse or the owner excessively. Our treatment utilizes intravenous oxytetracycline at a dose of 6.6 – 8.4 mg/kg once daily for 5 to 10 days and then continues treatment using a form of oral oxytetracycline for 2 additional weeks. Most horses (and their veins) can accommodate this treatment protocol <u>without</u> the use of intravenous catheters. Our success rate for achieving negative test results after treating is over 80%, still better than the results in the study using doxycycline and Naxcel. Reasons for the less than 100% success rate are threefold: re-infection (as horses return to their areas of infection and tick habitats); incomplete treatment (owners unable to or horses unwilling to complete the oral treatment); or incomplete resolution, but a greater than 50% reduction in circulating C6 levels (as the SNAP test only records color –although it can vary from faint to dark blue).

At the time of publishing, there is no commercially available approved test that measures the C6 level in a positive 4DX SNAP test. Tests are being conducted to determine if it is possible to measure a quantitative level of C6 in the horse. If successful, this will add more certainty to diagnosis and treatment success, as well as accuracy of determining re-infection.

As intravenous oxytetracycline chelates (or grabs) circulating calcium, it must be diluted and given slowly to ensure the heart and skeletal muscle do not have any problems. A typical treatment takes 3- 5 minutes. Also, it is not a treatment for untrained professionals as the antibiotic can be irritating to veins if not administered so that the entire dose is completely within the vein.

This protocol takes into consideration that intravenous oxytetracycline ONCE daily is superior to oral doxycycline or intramuscular Naxcel twice daily. This is due to higher tissue concentrations of oxytetracycline compared to low levels of oral doxycycline. Those that did respond to doxy or Naxcel in the Cornell study showed a significant decline in antibody level during treatment but also showed antibody levels increased after treatment was discontinued in 75% of each treatment group.

These two drugs appear to inhibit reproduction of, but not eradication of B. burgdorferi.

Doxycycline also has potent anti-inflammatory effects which may allow horses to show improvement in their lameness signs regardless of the drug's effect on the organism.

PREVENTION OF LYME DISEASE IN HORSES

Many different methods of prevention have been attempted and utilized. The recombinant DNA canine vaccine did show it was effective in preventing Lyme disease in ponies challenged with infected ticks 3 weeks after a three-dose regimen. The vaccine appears to inhibit the spirochete within the tick. This means as the tick ingests the blood meal from the host which is laced with the antibodies from the vaccine it will inhibit or kill the spirochete in the tick's hindgut.

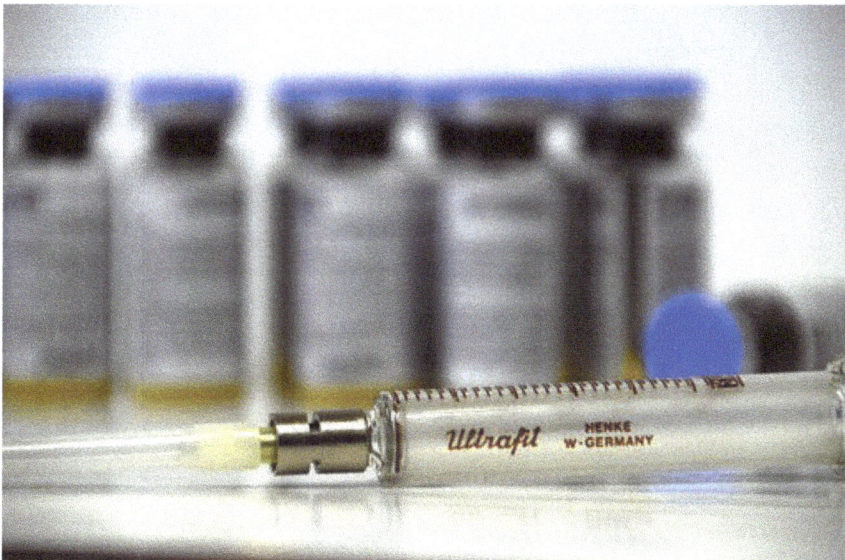

Therefore, ticks known to be carrying the spirochetes do not infect the host after engorging on a blood meal. Unfortunately, we do not how frequently to vaccinate and also do not know if it is safe to give to a horse which may have been previously infected or is currently not known to be infected. There are no safety studies. At this time we cannot recommend using a dog Lyme vaccine on your horse.

What we do know is that dogs bring ticks into homes and fields and our friends live in close proximity to people and horses. Dogs roll in leaves and run into tick habitats where the larval, nymphal, and adult stages lay waiting for their next blood meal. Dogs are reported to be 50% more likely to get Lyme disease than humans and are now used by human doctors to determine if Lyme disease is in the area. If a family member complains of symptoms similar to Lyme disease, quite often treatment is initiated if the family pet has also been diagnosed with the disease. Therefore, protecting pets often may all that be necessary to protect you and your horse. (Note: cats can get it too)

Horse prevention: Examine your horse on a daily basis; remove nymphs and adults immediately. Remember it takes at least 24 hours of attachment to transmit the disease. Be sure to check the neck, base of the mane, under the tail and the ears. If you see a tick, remove it by grabbing it as close to the skin as possible with fine tweezers. Pull straight up slow and steady. You can then put the tick in a sealed container (no alcohol) for testing. Once the tick is removed apply alcohol or antibiotic ointment to the site.

LYME DISEASE ALERT

DO A THOROUGH BODY CHECK FOR TICKS AFTER BEING OUTDOORS.

Deer tick size (left to right)
larva, nymph, adult

How To Remove A Tick

- Using tweezers, grasp tick near the mouth parts, as close to skin as possible.
  - Pull tick in a steady, upward motion away from skin.
  - DO NOT use kerosene, matches, or petroleum jelly to remove tick.
  - Disinfect site with soap and water, rubbing alcohol or hydrogen peroxide.
- Record date and location of tick bite. If rash or flu-like symptoms appear contact your health care provider immediately.

DISEASE RISK IS REDUCED IF TICK IS REMOVED WITHIN 36 HOURS.

2709   **New York State Department of Health**   8/02

Try to keep clear of tick-infested area. If not possible, use insect repellent with permethrins in spring, summer and fall– although none are approved for use on horses. As horses sweat, just like us humans, whatever we spray on them topically with dilute and ultimately become inactive. There are now permethrin wipes made for dog use that can be used for horses. We recommend using them before venturing into tick infected areas by wiping each leg, the main and tail once an hour while in the area.

Clean up leaf/brush piles in our yard, your paddocks, and on your farm. These are great areas for the white-footed mouse to live and continue the cycle of Lyme disease on your property. Mow your fields to again minimize the number of mice and also to minimize the vegetation ticks like. It is also recommended to use wood chips or gravel between lawns and wooded areas to restrict tick migration into your grass paddocks or yards.

White-footed mouse

Stack wood in dry areas where ticks are less likely to enjoy a safe haven. As we stated earlier, deer serve to spread the ticks wherever they go. Do not feed deer on your property and, if you need to, plant deer resistant plants to inhibit them from approaching too close.

White-tailed deer

You can also purchase or construct bait boxes to treat wild rodents with an acaracide (tick killer). These are available for home use in some feed supply stores or you can make them by placing treated cotton balls inside old paper towel or toilet paper rolls. Treat the cotton balls with permethrins or other topical tick products in order to have the rodents take the cotton balls back to their homes where they will make them part of the bedding and continuously rub up against them removing and/or killing any larval or nymphal stage ticks. This can help reduce ticks by more than 50%.

LOOKING TO THE FUTURE

The future looks to remove the source of infection, that is, the organism inside the white-footed mouse. A project was undertaken to vaccinate the mice to assess its effectiveness. 900 white footed mice in 12 different forested sites in CT were trapped and vaccinated with dog vaccine. The rodent's antibodies kill the bacterium inside the tick – preventing the tick from spreading the disease to the next host. After vaccinating 55% of the mouse population in targeted areas, researchers measured an overall reduction in the prevalence of Lyme disease transmission in nymph-stage ticks. As trapping and vaccinating mice is a never ending and labor intensive (and expensive) undertaking, the emphasis is now on the need to develop an oral vaccine (like Rabies). Initially, a large amount of money for funding the research to develop the vaccine is required. Then there will need to be trials conducted that measure the effectiveness in numerous different areas before a product will be available for use. This will take YEARS to develop.

There is <u>no</u> evidence of transmission directly from animal to human, and one cannot get infected from the urine of an infected animal. A horse also <u>cannot</u> get infected from eating a tick or portion of a mouse (in baled hay). The spirochete is not a highly resistant organism in the environment; it would probably be killed by the acidity in the stomach and small intestine and by the digestive enzymes. Plus, it must be introduced into the skin or directly into the bloodstream.

ANAPLASMA & EHRLICHIA

EHRLICHIOSIS is the 2nd most common canine infectious disease in the U.S. In 2001 Ehrlichia equi and Ehrlichia phagocytophila were reclassified as one organism called *Anaplasma phagocytophilum*. Co-infection with both Lyme disease and one of the other tick borne diseases is relatively common. One study done in dogs showed 45.9% of dogs testing positive for Lyme disease also tested positive for Ehrlichia. The human literature reports 9-26% of humans with Lyme disease are co-infected with other tick-borne illnesses (Anaplasmosis, Babesiosis, and Erhlichiosis). At our clinic 10% of horses infected with Lyme disease have also tested positive for Anaplasmosis.

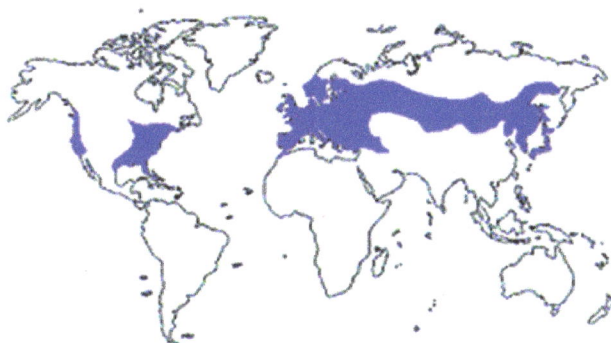

Distribution of Anaplasma worldwide

Both organisms are intracellular bacteria transmitted by ticks. They are small (0.2-1.0 um), intracellular, gram-negative organisms. They invade mammalian WBCs in which they multiply and form membrane bound, intracytoplasmic colonies called morulae.

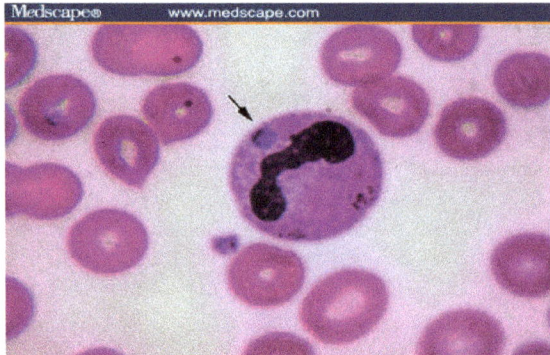

Morula

This means we can see them on a smear of blood cells under a microscope. *Ehrlichia canis* morula are found in large white blood cells (monocytotropic) and *Anaplasma phagocytophila* morula can be found in smaller white blood cells called neutrophils (granulocytic). After less than a week of infection, the morulae are often removed from the circulation as the body removes the abnormal white blood cells. The infection may remain, but the ability to diagnose it based on a simple blood smear is gone.

Anaplasma phagocytophila is found similarly to *Borrelia burgdorferi* in white-tailed deer and the white footed mouse. After infection of the horse, the incubation period is typically one week (range: 1 -21 days). The clinical signs develop over several days and include fever, swollen legs, depression, anorexia, depression or lethargy, and numerous joint soreness. Blood testing may show a decrease in platelet numbers.

DIAGNOSIS

As stated above a blood smear which contains the organism within white blood cells provides a definitive diagnosis, but as also stated above, they are commonly not seen due to timing of the infection.

4 DX SNAP Test:

The *Anaplasma phagocytophilia* portion of the 4DX SNAP test detects antibodies to Anaplasmosis species. This test is different than the portion testing for Lyme disease as it detects antibodies. It mimics a specific region of an outer membrane protein found on the *Anaplasma phagocytophilia* organism. This test is NOT subjective like an IFA test or titer, but does shows a good correlation with a positive IFA titer of 1:80. IFA tests are not specific tests for specific portions of antibodies and therefore may show cross reactivity with other types of Anaplasma and Ehrlichia. The SNAP test also detects both IgM (acute infection) and IgG (chronic infection). It can take 7 -21 days to mount an antibody response and any IgG may remain elevated for months to years. Therefore, negative tests may need to be retested in 1-3 weeks; and retesting after treatment may not be accurate in determining success of treatment.

The disease is self-limiting in most horses. Occasionally we will see small pin point area of platelet hemorrhages on the gums and inside lips. This is an indication of low platelet numbers. Rarely, this can progress to heart muscle and small vessel problems called myocardial vasculitis. Unchecked, this can cause irregular heartbeats called ventricular premature contractions ("VPCs"). Young animals have milder signs than older animals, often thought to be due to the immune strength of the younger animals.

There are 3 stages:

(1) <u>Acute</u>: brief, symptoms often missed. This is often confused with mild, passing viral infection (fever, lethargy), and a change in behavior. Achieving a cure is best now with treatment OR self resolution removes the infection. Testing may be useless as it is often too soon for antibody titers to rise. Therefore, many of these infections are often missed.

(2) <u>Subclinical</u>: no symptoms or containment and/or resolution of infection by the horse. There may or may not be platelet count decreases, as well as increased antibody titer. These horses appear abnormal only briefly and then return to appear normal.

(3) <u>Chronic</u>: Infection and signs last several weeks to months. There are often severe symptoms and often the infection was dormant and has now reappeared for some reason. Clinically, we often see low platelet counts (Thrombocytopenia) which can cause abnormal amounts of bleeding for what appears to be small issues. If left unchecked and untreated, the thrombocytopenia can progress to the bone marrow leading to further problems as other blood lines become altered.

TREATMENT OF ANAPLASMOSIS AND EHRLICHIOSIS

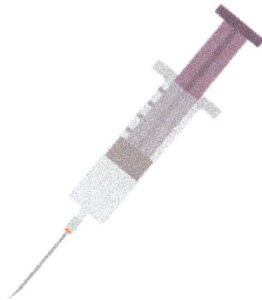

When suspicion or testing confirms an infection, the disease can be self-limiting and eliminated by self-resolution, or it can be treated with antibiotics. The standard antibiotic treatment is intravenous oxytetracycline once daily for 5 to 8 days, or oral doxycycline for 7 to 10 days. The most striking change or improvement noted by owners is typically the behavior; the horse's attitude improves. Clinically, we see a slight increase in platelet numbers. Commonly, we also utilize non-steroidal anti-inflammatory drugs (NSAIDs) for the fever and soreness the horse maybe experiencing

PREVENTION OF OTHER TICK BORNE DISEASES

Prevention of these diseases is similar to Lyme disease prevention.

- Remove ticks within 24 hours

- Clean up wood piles and brush

- Sprays or wipes before going into heavy tick areas

- Treat your dogs and cats

- By protecting your pets will help to protect yourself and your horse

❊

Tick borne diseases are on the rise all over the world. New strains of Borrelia organisms, new geographical patterns, and new treatments are being reported commonly. New research continues to bring out new information on how to diagnose, treat, and prevent these diseases. Lyme disease can be a very frustrating dilemma for you and your horse. If you have any questions, please do not hesitate to ask your veterinarian or give my office a call.

❊

REFERENCES

Manion TB, MS; Bushmich SL, Mittel L, Laurendeau M, Werner H, and Reilly M, Lyme Disease in Horses: Serological and Antigen Testing Differences, AAEP Proceedings 1998, 144-145.

Straubinger RK, Lyme Borreliosis In Dogs, Recent Advances in Canine Infectious Diseases, Carmichael L. (Ed.) International Veterinary Information Service, Ithaca NY, 2000.

Divers TJ, Chang YF and McDonough PL, Equine Lyme Disease: A Review of Experimental Disease Production, Treatment Efficacy, and Vaccine Protection, AAEP Proceedings, 2003.

Chandrashekar R, Daniluk D, Serologic Diagnosis of Equine Borreliosis: Evaluation of an In-Clinic ELISA (SNAP® 3Dx®), Idexx Laboratories, 2004.

S. M. McDonnell, Is it Psychological, Physical, or Both? AAEP Proceedings, 2005.

Alleman AR, The Diagnosis and Treatment of Tick Borne Diseases in Dogs, NAVC Proceedings 2005, 472-477.

Wilson, JH, Vaccine Efficacy and Controversies, AAEP Proceedings 2005.

Straubinger RK, What Are Ticks Doing These Days? Emerging Diagnostic Research and Vaccine Management Considerations, NAVC Proceedings 2006.

R.E. Goldstein, Borrelia Burgdorferi, Lyme Nephritis and Total Lyme Disease Management, NAVC Proceedings 2006.

Littman MP, Goldstein RE, Labato MA, Lappin MR, and Moore GE, ACVIM Small Animal Consensus Statement on Lyme Disease in Dogs: Diagnosis, Treatment, and Prevention, J Vet Intern Med 2006;20:422–434.

M.P. Littman, Lyme Disease: What to Do When the Snap Is Positive, NAVC Proceedings 2006.

Johnson AL, Divers TJ, Chang YF, Validation of an in-clinic enzyme-linked immunosorbent assay kit for diagnosis of Borrelia burgdorferi infection in horses, J Vet Diagn Invest 20:321-324 2008.

Lisa A. Fortier, Current Concepts in Joint Therapy, Proceedings WEVA 2009.

Goldstein, R.E., Infectious Diseases and the Kidney, Proceedings WSAVA 2010.

Schnabel LV, Papich MG, Divers TJ, Altier C, Aprea MS, McCarrel TM, Fortier LA, Pharmacokinetics and distribution of minocycline in mature horses after oral administration of multiple doses and comparison with minimum inhibitory concentrations, Equine Vet J. 2012 Jul;44(4):453-8.

The Importance of Differentiating Exposure from Infection with Borrelia burgd orferi in the Diagnosis and Treatment of Canine Lyme Disease, Idexx Reference Labs, 2012.

ABOUT THE AUTHOR

Mark T. Reilly, D.V.M, Diplomate ABVP (Equine) is originally from the South Shore area of Massachusetts, having graduated from Abington High School. After attending the University of New Hampshire and earning a B.S. in Animal Science, he went on to Tufts University School of Veterinary Medicine. After graduating from Tufts in 1991, Dr. Reilly set off to work on the East coast horse racing circuit spending time working in the New York/New Jersey area and winters in southern Florida. He then moved on to work at Delaware Park, before landing on Cape Cod where he ultimately established a large animal ambulatory practice in 1995. For the next 10 years, Dr. Reilly traveled to Nantucket, Martha's Vineyard, Cape Cod, and southeastern Massachusetts for farm calls throughout most of the year. Winters were again spent in Florida working on young racehorses in training, as well as on horses competing at the Wellington Equestrian Festival.

After a few years of searching for a location to establish an equine hospital, construction began in the spring of 2005 in Plympton. The South Shore Equine Clinic & Diagnostic Center opened its doors in January 2006.

South Shore Equine Clinic & Diagnostic Center is located in Plympton, Massachusetts, in the southeast area of the state, 30 miles south of Boston and 15 miles north of Cape Cod. The full service equine medical and surgical center also provides ambulatory field service and treats horses ranging from the family companion to the race horse and show champion. In addition to wellness and senior care, the clinic's veterinarians provide a wide range of services including advanced dentistry, ultrasonography, endoscopy, reproduction and neonatal services, sports medicine, acupuncture therapy, soft tissue and orthopedic surgery, internal medicine and ophthalmology. The clinic sees patients from throughout southeastern Massachusetts, Cape Cod and the neighboring islands, as well as receives referrals from the New England area.

Dr. Reilly is certified as a Diplomate by the American Board of Veterinary Practitioners (ABVP) specializing in Equine Practice. ABVP Diplomate status is granted under the approval of the American Board of Veterinary Specialties, an official committee of the American Veterinary Medical Association. While other veterinary specialties focus on specific disciplines or organ systems (e.g. ophthalmology, cardiology, etc.), ABVP diplomates demonstrate excellence in all areas of specialty care of the total patient. Currently there less than 90 veterinarians that are certified Equine ABVP diplomates in the United States, including Dr. Reilly. He is the only ABVP certified equine private practitioner in New England at the time of publishing.

Dr. Reilly's has spent over 25 years working on competitive horses of all disciplines. His professional interests are pre-purchase evaluations, lameness/poor performance issues (feet/shoeing, cardiac issues, joint injuries, airway endoscopy, IRAP, Stem Cell, PRP), reproduction (artificial insemination—fresh cooled and frozen semen, reproductive tract evaluations, and pregnancy/pre-natal issues), LASER surgery, respiratory disease, diagnostic imaging (x-rays, gastroscopy, ultrasonography, MRI), preventative health care, and internal medicine. He has been president of the Cape Cod Veterinary Medical Association, as well as chairman of the Massachusetts Veterinary Medical Association Large Animal Committee. Dr, Reilly is a lecturer at the Tufts Cummings School of Veterinary Medicine in North Grafton, MA, and at the Integrative Veterinary Medical Institute in Reddick, FL.

Dr. Reilly presented numerous papers at the prestigious international convention for equine practitioners (AAEP). He has presented on a novel castration technique at the 2005 international convention, a technique for treating respiratory disease in 2012, and numerous articles on business principles at the 2009, 2010, 2011, 2012, and 2013 AAEP Annual and/or Summer Focus conventions. In addition, he has lectured and led a wet lab on the use of Platelet Rich Plasma (PRP) at the Hambletonian CE Seminar in 2014 and was the invited Large Animal Distinguished Coughlin Professor at the 2017 University of Tennessee, College of Veterinary Medicine Annual Conference for Veterinarians and Veterinary Technicians (The Coughlin Symposium).

Dr. Reilly also enjoys lecturing to owners and trainers at local and regional meetings. When not spending time with his family and friends, he is an avid sports fan and enjoys working in his yard, spends time at the beach and traveling.

CPSIA information can be obtained
at www.ICGtesting.com
Printed in the USA
LVHW050910181219
640672LV00014BA/827/P

9 781641 514682